HOME SPACE CREATIVE DESIGN ROUNDUPS +
居家空间创意集

田园休闲
RURAL LEISURE

深圳市海阅通文化传播有限公司 主编

中国建筑工业出版社

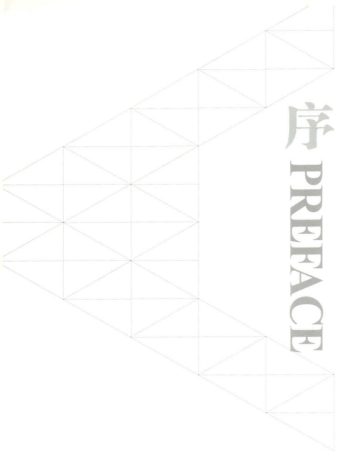

序 PREFACE

Do you want to own a land of idyllic beauty belonging to yourself? Do you want to have a harbour to rest your heart and soul? Do you want to wander about in a fresh and natural space or to listen to nature, get close to nature and return to nature? If your answer is YES, then I believe that this book will surely bring your unexpected surprises.

This book collected masterpieces from many excellent designers. And these works expound the phrase Idyllic and Leisure. They gain popular favor though not elegantly carved and carefully chiseled, no carved beams and painted rafters nor dazzling luxury. The reason lies in that Idyllic Style advocates " Returning to nature", which can exactly make people now living in such a high-tech and fast-paced society strike a balance both physically and mentally. Hence, Idyllic Style wins growing popularity.

Idyllic style adopts natural materials, such as bricks, pottery, timber, stones, rattans, bamboos and green potting. In terms of the textile material, cotton and hemp fabrics are preferred, texture of which corresponds to the rustic-style pursuit of nature with little adorning. It is the idea and selection of materials that creates a natural, plain and elegant atmosphere.

It is really relaxing and pleasant to enjoy a cup of tea and exchange funny stories with friends in fresh idyllic atmosphere.

想拥有属于自己的"世外桃源"吗？想拥有心灵憩息的港湾吗？想徜徉于一片清新自然的空间吗？想聆听自然，亲近自然，回归自然吗？如果答案是"YES!"，那么我相信，这本书会带给你意想不到的惊喜！

本书汇集了各位优秀设计师的杰出作品，这些作品把"田园休闲"四个字阐述得淋漓尽致，尽管没有精雕细琢，没有雕梁画栋，更没有"贵"气逼人，但是却赢得了大众的青睐，原因在于田园休闲风格倡导"回归自然"，而这种自然恰好能让人们在当今高科技快节奏的社会生活中获取生理和心理的平衡，因此田园休闲风格愈来愈受到大众的青睐。

田园风格采用自然的材料，如砖、陶、木、石、藤、竹、绿化盆栽等，织物质地上也选择棉、麻等天然制品，其质感与乡村风格不饰雕琢的追求相契合，也正是这种理念和材质选择，才塑造了一种自然、简朴、高雅的氛围。

在清新的田园氛围中 和朋友一起品一杯香茗，聊聊趣事，实在放松。

本书精选了多个极具代表性的作品，多方面展示了简约
欧式风格的特点。设计师们在此风格上的匠心独
运，既满足了国人的品位，又展示了现代
流行。

居家空间创意集
HOME SPACE CREATIVE DESIGN ROUNDUPS

田园休闲 RURAL LEISURE / 目录 CONTENTS

04 Nanjing American-style Villa
南京美式别墅休闲生活

10 Shenzhen Xiangmi Lake NO.1, 17B Private House
深圳香蜜湖一号 17B 私人豪宅

16 Suntech City-state A Seat Example Room
尚德城邦 A 座样板房

24 Shen Ye Peng Ji Huizhou Wanlinhu 6 Period Example Room
深业鹏基惠州万林湖 6 期样板房

30 The Historic House in Summit NJ Today
历史私人住宅

36 Let The Sunshine Dance in the Home
让阳光在家中飞舞

44 The South of Colorful Ⅰ
彩云之南 Ⅰ

50 The South of Colorful Ⅱ
彩云之南 Ⅱ

54 Wonderful Place
优岙美地

62 Quiet Life in the Garden
花园中宁静生活

68 British Idyllic Style
英式乡村风

74 Swan Castle
天鹅堡

82 Zhongnan Century City Building 5
中南世纪城 5 幢

Home Space Creative Design Roundups 居家空间创意集

NANJING AMERICAN-STYLE VILLA
南京美式别墅休闲生活

Desginer: Liao Xinyao　　　**Location**: Jiang shu　　　**Area**: 320m²
设计师：廖昕曜　　　　　　　项目地点：江苏　　　　　　面积：320 平方米

This is a luxury villa, the area is 320sq meters. It locates in XianLinShanShuiFengHua housing estate which is surrounded with beautiful scenery. After redesigned and renovated, entrance door and master room turns to south, hallway is enlarged, half floor is underground while three floors are overground. Ventilation and lighting are very good, space is square and upright, layout is reasonable, ceiling is raised. The interior is generous and cultural.

The program gets rid of complication and luxury, using comfortable function as guide,emphasizing natural relaxation.American furnitures, rustic tiles, large textile sofa and floral curtains, the color of them are all elegant and mild.Classical elements and animal patterns are added to decorations, creating a desirable peace.

Layout of function mainly consists of recreation,relaxation,sleeping and living. Every room becomes more spacious and brighter.You can reading, drinking a cup of tea,playing here as you like.

White ceiling reflects a concise and natural feeling.Beige and white wall with some decorations shows a plain and elegant sense.

In the inner room, beams over head, sunlight shined the table, simple bookshelf and articles for daily use reflect cozy and tranquil life of owner. A strong flavour of life takes you to an American manor,and bring you a natural breezing.There are no luxury materials, dazzling lights,gorgeous color,but we can see dignity in nature.Interior forms a harmony with outside.It satisfies modern people's attitude for nature and fashion for simpleness and low carbon.The design shows a natural, harmonious, full picture.

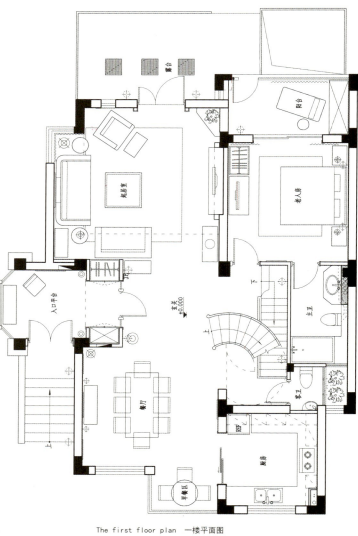

The first floor plan 一楼平面图

本案例属于别墅豪宅设计类别，面积为320平方米。别墅地处仙林山水风华小区，自然环境优美。此案经过重新设计改造后，入户大门朝南，主卧朝南，门厅加大，空间上分设地下半层、地上3层，通风采光好，户型方正，结构合理，空间高，而内部的设计使整个别墅既落落大方又不失文化底蕴。

本案摒弃了繁琐和奢华，以舒适机能为导向，强调自然休闲。美式家具、仿古地板，宽大的布艺沙发，碎花窗帘，色彩典雅温和。配饰加入古典元素、动物造型，营造出一种令人向往的舒适宁静。

功能布局主要以休闲、休息、睡眠、起居为主，经过改造，每层空间都变得宽大明亮，休闲，阅读，品茗，弹奏，无所不可。

整个色调以黄白咖啡色相间，白色的顶棚，显得简洁自然，墙面运用一些米黄，与白色相间，加以一些饰品装饰，让整个空间充满质朴、高雅之感。

Rural Leisure 田园休闲

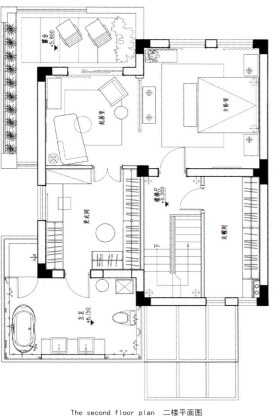

The second floor plan 二楼平面图

步入内室，头顶的木梁、照在古老餐桌上的那抹阳光、床边简易的书架，随意摆放的生活用品，无不折射出主人生活的惬意与恬淡，浓郁的生活气息扑面而来，仿佛把人带到美式庄园，给处于现代喧嚣闹市中的人吹来一股大自然的清新之风。在这里，虽然没有昂贵的材质、耀眼的灯光、绚丽的色彩，但古朴中不乏尊贵，室内、室外遥相呼应，符合现代人返璞归真的心态和追求简单、低碳的生活潮流，整个设计显得十分自然、和谐、完整。

SHENZHEN XIANGMI LAKE NO.1 ,17B PRIVATE HOUSE
深圳香蜜湖一号 17B 私人豪宅

Designers: Zheng Hong, Xu Jing
设计师：郑鸿 徐静

Area: 350 m²
面积：350 平方米

Photography: Xu Jing
摄影师：徐静

New customs are always particularly impressed by the colorful world when they visited this house. They can abandon the noisy and immerse themselves in a queit, leisurely world, as if they lived in a ideal space far away from realistic society.

They were surprised by the fact that the owner of the house is an eighty-years old couple. The concept that aged people should live solemn house with colorless started to shake. Will a colorful space make them comfortable and happy?

我带一些新客户去参观这套房子的时候，他们都暗自被这个色彩斑斓的世界所折服，仿佛置身于空中楼阁，外面的那些喧嚣、浮华与这里毫无关联，有的只有宁静、淡然、闲适。

当他们得知房子的主人是两位已年近八旬的老人时，更是大为惊叹：为什么以前的观念里都一直认为老人的房子就应该是老成稳重的呢？让老人住在这般缤纷多彩的空间里，岂不是让他们倍感舒心与愉悦？

Rural Leisure 田园休闲

The old couple spent more than half of their lives on Shanghai and moved to Shenzhen in their old age. Considering this fact, I chose furnitures of Shanghai style, an old tea table and rosewood cabinet to add their familiar atmosphere in a new city.

The diamond-shape floor can be found everywhere. The house is decorated with most Shanghai style furnitures on the premise of rural style.

考虑到两位老人都是典型的上海人，在上海度过了大半辈子，直至晚年才移居深圳，在设计时，更是特意配上了海派家具：一张老上海式的茶几，一个略带厚重感的红木电视柜，好让他们在这个也许还算陌生的城市体味到那份独属于家的温情感。

整套房子贯穿始终的都是格菱纹状的地板，在以田园风格为主的前提下，点缀了不少海派家具。

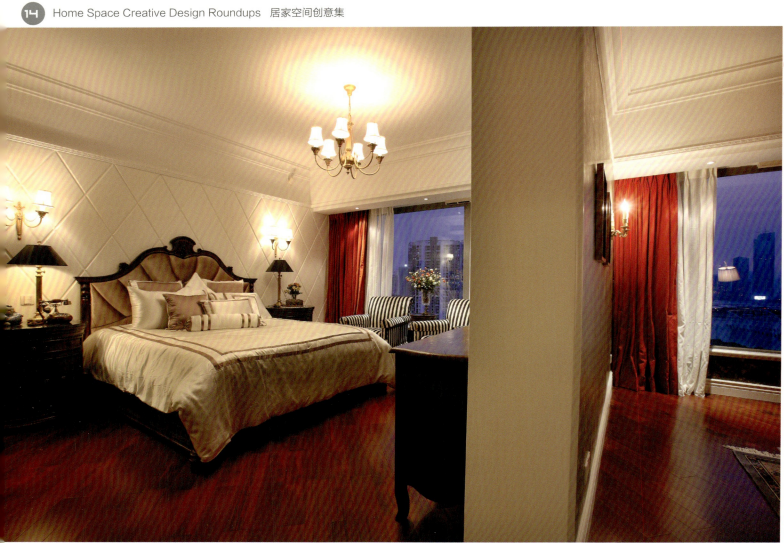

Home Space Creative Design Roundups 居家空间创意集

SUNTECH CITY-STATE A SEAT EXAMPLE ROOM
尚德城邦 A 座样板房

Design company: Xing Yu Tian Design Company
设计公司：行于天设计公司

Location: Natong, Jiangsu
地点：江苏南通

Area: 150m²
面积：150 平方米

On a mini road of a little yard with fragrant odor, the flowers get a bit drunk, shadows of the moon is waning, the time touches here quietly like the water. With the memories of some long recovery brewing, a soul moving is being aroused, the secret of happiness is being wantonly enjoy. Pulsating with a distant hirtorical story but never abandone the life passion. Encompassed with fresh and elegant nature but never estrange from the bustling city. Here, when you standing on the breeze with wine and singing under bright moon, you will find a vibrant life moving up and down surrounded by the long years of reincarnation.

小园香径、群花微醉，月影阑珊，时光如水，静静抚过这里。某种久远复苏的记忆正在酝酿，灵魂深处的感动正在激越澎湃，心底的快乐正被肆意放逐。承继辽远的历史脉动，却从未摒弃生活的激情；坐拥自然的静雅清灵，却从未疏离都市的繁华。在这里，把酒临风、对月当歌，在悠悠岁月的变幻轮回中，生命的绿意处处涌动……

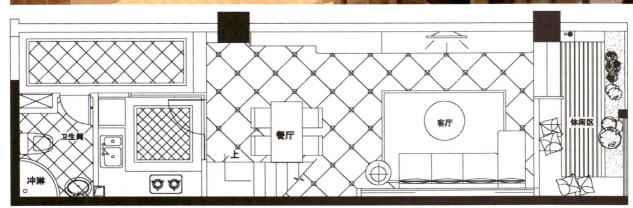

The first floor plan　一楼平面图

Rural Leisure 田园休闲

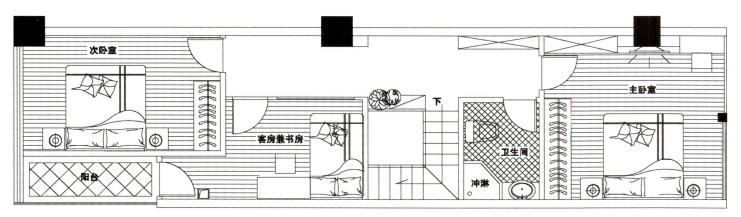

The second floor plan　二楼平面图

Rural Leisure 田园休闲

SHEN YE PENG JI HUIZHOU WANLINHU 6 PERIOD EXAMPLE ROOM
深业鹏基惠州万林湖6期样板房

Designer: Qu Shan　　**Materials**: compound wood floor, cream-colored texture painting, wall paper, white oak, rustic tile, dark brown mirror. etc.
设计师：屈杉　　装饰用材：复合木地板、米色肌理涂料，墙纸，白橡木、仿古砖及茶镜等

This is a model design of 75 sq.m two-bedroom in the sixth phase WanLinHu, Huizhou, ShenYePengJi. These rooms aims at young couples under 30 years old. The exquisite, romantic characters and the affordable price satisfied young couples.

The design bears a idyllic style. The living room is concise. The plain ceiling with horn line and wooden floor is natural without any unnecessary decorations. The dining table cabinets and the TV wall are perfect harmony with the spirit of nature. The white wainscot, dark brown mirror and decorative plates set a cosy, idyllic style in a mild, cream-colored atmosphere.

The layout of dining room is reasonable. The generous ceiling and the floor, elegnat wallpapers, rustic furnitures, texiles and decorations creat a sweet, romatic family.

本案为深业鹏基惠州万林湖项目6期的75平方米的小两房样板房设计。此房型的目标业主多为30岁以下的年轻夫妇，小巧精致、温馨浪漫、造价适中是他们普遍的要求。

本案定位于田园设计风格，客厅装饰质朴简练，角线平顶顶棚及木质地板，无任何多余装饰，自在洒脱。一组造型把餐桌壁柜、电视背景与墙面关系巧妙融合在一起，回归质朴，无矫揉造作之感。白漆护墙板、茶镜及壁挂装饰盘，衬托在温馨柔和的米色基调氛围内，定义了舒适闲暇的田园风情。

卧室区域功能配置合理，顶棚地面装饰大方得体，墙面选用清新浪漫的壁纸装饰，配以乡村气息浓郁的家具、布艺及配饰品，共同演绎出了小家庭温馨浪漫的幽雅情怀。

Rural Leisure 田园休闲

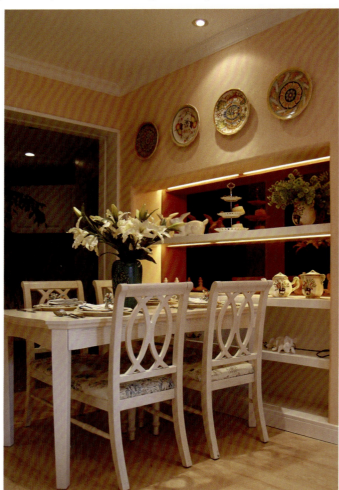

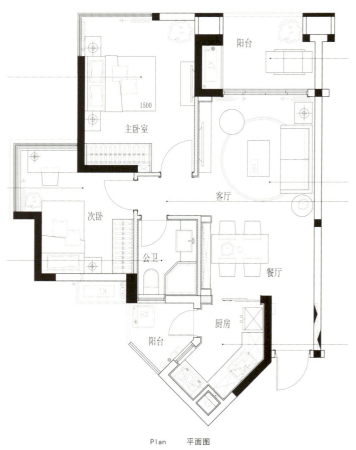

Plan 平面图

Rural Leisure 田园休闲

Rural Leisure 田园休闲

THE HISTORIC HOUSE IN SUMMIT NJ TODAY
历史私人住宅

Design company: Jo Ann Stephens Alston   **设计公司:** Jo Ann Stephens Alston

A previous owner had removed the original French doors and installed sliding glass doors inconsistent with the elegant architectural details of the room. Using hand blown glass, antique cremone bolts, and reproduction moldings, the designer created new French doors and sidelights that match those in the dining room. The room is furnished in a combination of English and French antiques that are in perfect harmony with the spirit of the home.

为了与室内精巧、雅致的建筑细节保持风格一致，主人用滑动玻璃门替代了双扇落地玻璃门。吹制玻璃工艺品，古典门栓，仿古摆件，设计师用全新的双开玻璃门和柔和的侧灯将这一切有机地结合，恰到好处。英式风格和法国古董完美融合，整个空间充满家的气息，和谐美好。

Rural Leisure 田园休闲

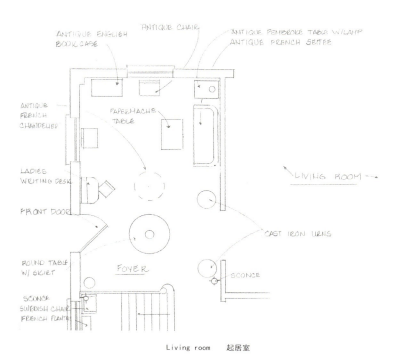

English garden floral's, a recurring element throughout the home, appear here in the master bedroom, too. The tie-back drapery treatment cleverly disguises the fact that one window is actually a French door that leads to the second floor porch. The ceiling is painted a soft lingerie peach that reflects the colors of the bed treatments, an antique tablecloth is used as a coverlet on the bed.

英式花园元素贯穿整个空间，也同样出现在主卧室中。浪漫神秘的窗纱，遮掩着后面的风景，似乎是一扇窗若隐若现，其实是扇门，带我们带来到二层的门廊。顶棚柔和的透明桃色折射出床上用品的色彩情调，床罩也采用了具有古典韵味的桌布。

Living room　起居室

Rural Leisure 田园休闲

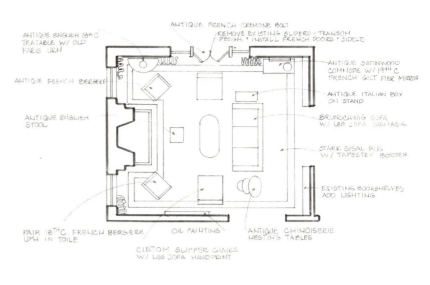

Family room 家庭房

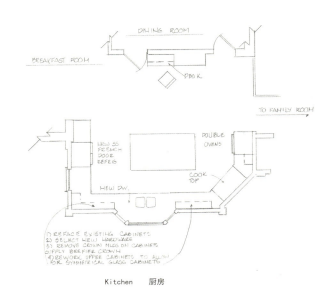

Kitchen 厨房

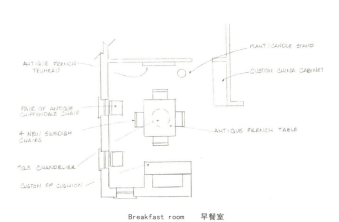

Breakfast room 早餐室

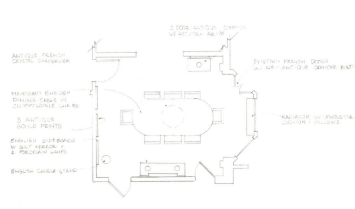

Dining room 餐厅

Rural Leisure 田园休闲

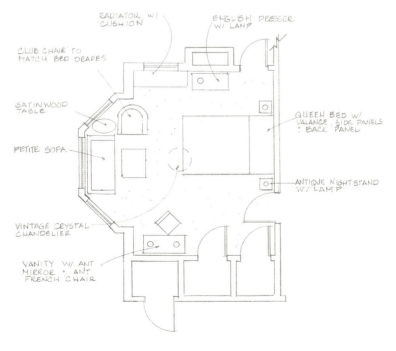

Master bedroom　主人卧室

Home Space Creative Design Roundups　居家空间创意集

LET THE SUNSHINE DANCE IN THE HOME
让阳光在家中飞舞

Design company：Xiong Longdeng Design Studio
设计师：熊龙灯工作室

Location：Beijing
项目地点：北京

Photography：Yun Wei
摄影师：恽伟

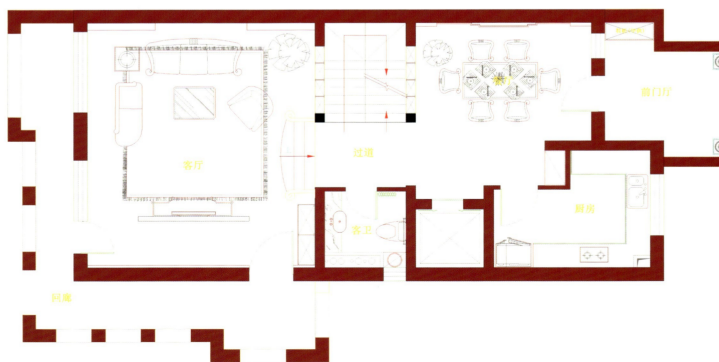

The first floor plan　一楼平面图

The townhouse located in Xiaotangshan bears a Spanish garden architectural style in. The house has five floors with three overground, two underground and an elevator. The ground floor is embraced by garden from three sides, we can enjoy green and sunshine at any place of the ground. But, the layout seems a little crowded. The second floor underground connects the public garage. It has a SPA area and an equipment area, we can also enjoy ourselves with LongMai hot spring here. Shower rooms are set in the second and third floor. The first floor underground is for recreation, video, cellar, tea house and sunshine yard. It links the family, also a good place to cultivate minds.

Rural Leisure 田园休闲

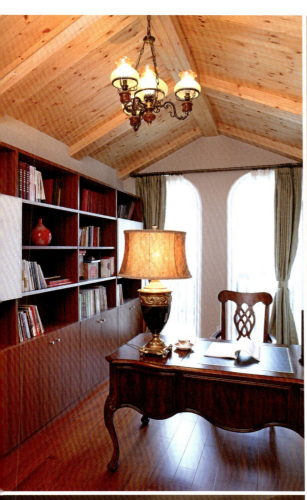

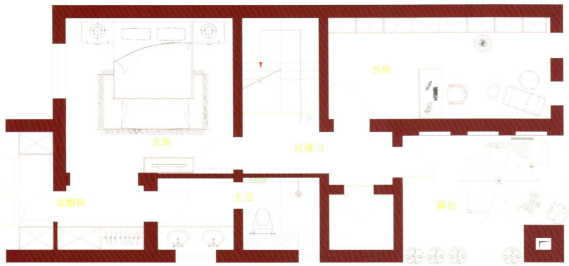

The second floor plan　二楼平面图

Home Space Creative Design Roundups 居家空间创意集

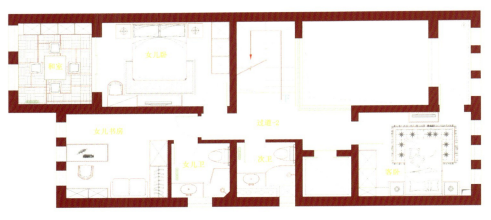

The third floor plan 三楼平面图

Rural Leisure 田园休闲　41

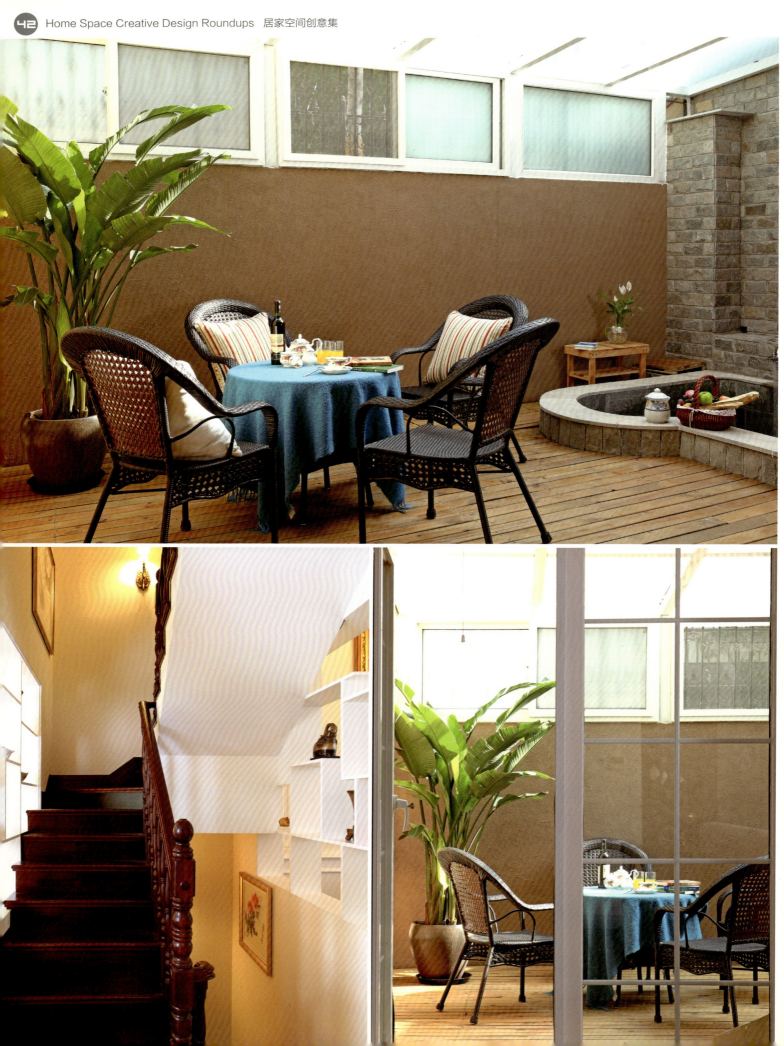

Rural Leisure 田园休闲

位于小汤山的联排别墅,为西班牙庭院建筑风格。房子5层,地下2层,地上3层,带电梯。房子优点在于地面一层三面回廊环绕,可以在室内任意空间通达室外庭院,感受夏日绿意盎然的庭院之美。缺点是室内格局小而密,空间比较紧凑。地下二层连接公共车库,功能分为SPA区和设备区,充分享用小汤山龙脉温泉;二、三层卫生间设淋浴房,节省空间;地下一层为家庭娱乐休闲区域,影音室、酒窖、茶室、下沉式阳光庭院,是家庭的纽带,也是修身养性的好去处。

THE SOUTH OF COLORFUL I
彩云之南 I

Designer: Liu Yaocheng
设计师：刘耀成

Location: Changsha
项目地点：长沙

Area: 135 m²
面积：135 平方米

The room decoration is a typical idyllic style. The designer want to show us a refinement house based on life.The interior is full of nature and sweetness.It can arouse our longing for freedom and outlook on life from heart.

本案例是一个典型的田园休闲风格的空间，在设计师的精心打造下，凝练成了这个以生活为本的精致居所。走进去，一股小清新的甜美之风扑面而来，整个设计充满了大自然的气味，也唤起了内心深处对自由的向往，对生活的展望。

Plan 平面图

THE SOUTH OF COLORFUL II
彩云之南 II

Designer：Liu Yaocheng
设计师：刘耀成

Location：Changsha
项目地点：长沙

Area：135m²
面积：135 平方米

Green is a kind of magic color,it relaxs us,making us feel the breath of lives.This program uses different green exquisitely to creat a colorful and orderly world, simple and profuse.Besides,lights are sophisticated,they light our vision up exactly right.This is a subtly collocation in space and can be used for references.

绿色是一种神奇的颜色，总能让人放松，且感受到生命的气息，这个设计恰到好处地引用不同层次的绿色，装点出了一个缤纷但又一点都不觉得眼花缭乱的世界，简单而不单调，丰富而不累赘，这正是设计师的精心搭配。除了颜色的搭配和空间造型，灯的运用似乎也是相当讲究，不多也不少，不偏不倚，正好照亮了我们的视线和想看到的效果。这是一个微妙空间的搭配，很有借鉴意义。

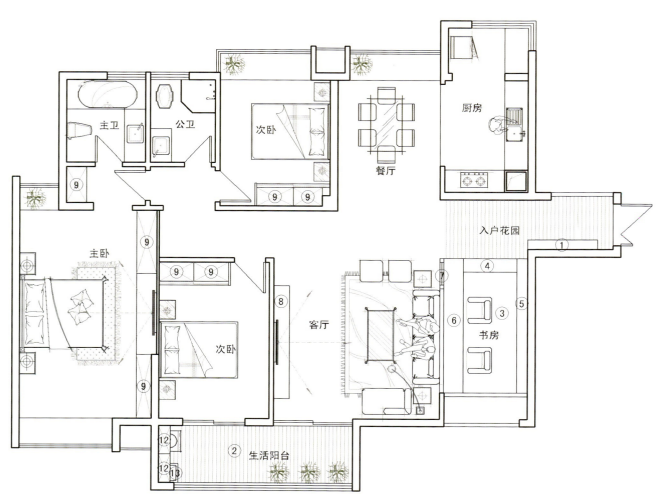

Plan 平面图

Home Space Creative Design Roundups 居家空间创意集

WONDERFUL PLACE
优恋美地

Design company：Dao Huo Niu Design Company	**Location**：Shenzhen Vanke City	**Area**：300m²
设计公司：导火牛	项目地点：深圳万科城	面积：300 平方米

Rural Leisure 田园休闲

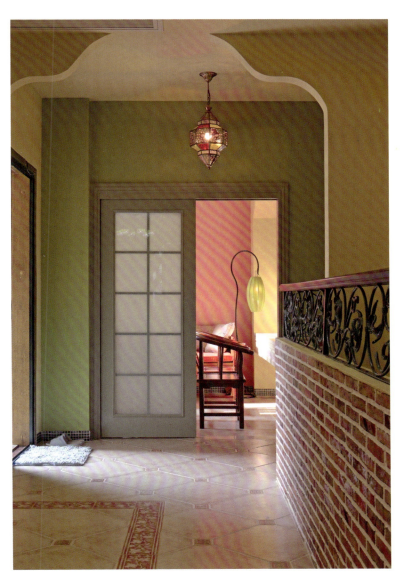

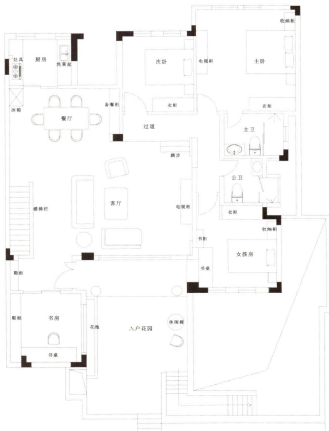

The first floor plan 一楼平面图

Home Space Creative Design Roundups 居家空间创意集

The owner had a concept of classic fitment and always paid too much attention to shapes. But designer's idea is "less fitment, more decoration". We must change Square idea. The question about how to choose the style didn't solve until bought the first group of furnitures. Then designer knew that idyllic style is right. The owner has been very busy.

So designer designed all work indepently. Without disturbance, as consequence, the unique idyllic style house was born.

如果老是整造型，这和"轻装修，重装饰"的设计思路有冲突，设计师必须找合适的机会完全颠覆业主那些顽固不化的硬装观念。直到选了第一批田园风格家具后，"受干扰"的设计思路才渐缓过来。每次挑到合适的灯具和配饰，拍完照片后设计师就打电话给业主，但是每次业主都由设计师决定。

于是自由的发挥空间塑造出了不一样的田园风。

Rural Leisure 田园休闲

The basement plane design 地下室平面设计图

QUIET LIFE IN THE GARDEN
花园中宁静生活

Design company: Yi Kuai Design Company
设计公司：壹块设计馆

Location: Shenzhen
项目地点：深圳

Area: 168 m²
面积：168 平方米

Rural Leisure 田园休闲

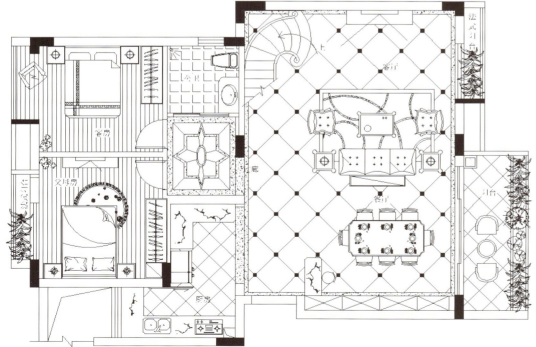

The first floor plan 一楼平面图

It seems like that you came to a serene garden from the nosiy city when you went to the construction site. The garden is consist of high-rise buildings around and the center houses. When you go through the high, depressive building, walking along the garden path, crossing the small brige, and reaching your white house in mediterrcnean style, you will feel so nice being back home.

The style is expected to be cozy, splendid with a little romatic feeling. At first, the young couple wanted the European style only. Now, it become more aboundant with some rural and modern elements. It also brings the couple a more romantic and sweet home.

每次去施工现场都感觉像从喧嚣的城市慢慢走向宁静花园；花园是由周边的高层建筑和中心的洋房构成，通过周边压抑的高层建筑，慢慢地迈着小步、踏着小路、过着小桥沿着小溪来到中心的洋房，看着地中海式的一幢幢白色洋房，突然感觉"回家真好"……

风格就这样慢慢敲定下来，在设计中尽量体现舒适、华丽而夹带一点点小浪漫。体现整体的欧陆风格外另加入一点点田园与现代的元素，使原有欧式不那么单调，最主要业主也挺喜欢，原本他们只提出要欧式风格，加上点田园与现代元素使这对年轻夫妇多了一些浪漫与现代感，多了一份家的温馨……

Rural Leisure 田园休闲

The second floor plan　二楼平面图

BRITISH IDYLLIC STYLE
英式乡村风

Design company: Taoxi Space Design Office
设计公司：陶玺空间设计事务所

Location: Taibei
地点：台北

Area: 119 m²
面积：119 平方米

In order to remain parts of former furnitures and rose patterns, the design uses a Birtish idyllic style to present a tranquil and elegant space.

Hallway——According to western and taking advantage of layout. A living room combining study and piano room is set here as a buffer rest space.

Living room——the wallpaper with floral patterns on TV wall stretches to dining room, adding an elegant feel of idyllic style. All the appliances are integrated in a single large marble block. There are not too much cabinets that expands interior visual space.

Dining room——The open kitchen expands the space, rustic floor tiles create areas. The bar also has the function of a countertop. White kitchenwares and mosaic wall will give you a happy cooking day.

Master bedroom——Carpet and curtain with rose patterns, silver wallpaper, and canopy bed with sheer. A clear beautiful feeling emerges under all of them.

Girl's room——This room uses pink as the elementary tone. The floral wallpapers and butterfly curtain make room a pincess castle. Lights and shadows create a sweet and romantic atmosphere.

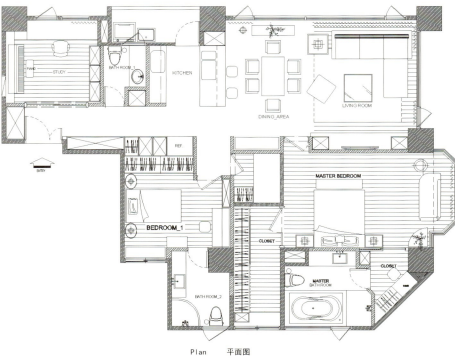

Plan 平面图

为了保有业主原有部分家具,加上业主喜爱的玫瑰图腾,全室以英式乡村风呈现恬静优雅的居家空间。

门厅 顺应格局优势,参考欧美居家配置,在门厅附近设置了融书房、钢琴房等于一体的多功能室,作为一个休憩的缓冲空间。

客厅 电视主墙的花纹壁纸延伸至餐厅,衬托出乡村风典雅的气氛,大理石台面将电器线路隐藏收纳,且没有过多的柜体,放大了室内的视觉空间。

餐厅 开放式的设计让餐厨空间放大,以复古地砖区隔出空间界定,吧台兼具料理中岛的功能,纯白厨具搭配缤纷的马赛克墙面,使得料理做菜的心情更加愉悦。

主卧室 玫瑰图腾的地毯及窗纱、淡雅的银白色壁纸墙面、四柱床装扮上浪漫布幔,莹白洁净的美感应运而生。

女孩房 以粉嫩色调为主,小花壁纸及蝴蝶窗纱将女孩房布置成公主城堡般,从窗纱中隐约透露的光影,营造出小女孩的甜美浪漫氛围。

SWAN CASTLE
天鹅堡

Design Company: Shenzhen Yi Pai Interior Design co., LTD
设计公司： 深圳市伊派室内设计有限公司

Designer: Duan Wenjuan
设计师： 段文娟

Location: Shenzhen
项目地点： 深圳

The Swan Castle locates Overseas Chinese town, Shenzhen with lakes and hills surrounded. Close to nature is the design principle of this project. There are four families in every floor. The living room is throughout north and south and family can overlook the beautiful swan lake in the center of yard easily. We try our best to express natural, idyllic life and create a cozy, relaxed home for you. You can enjoy the beautiful nature just at home!

Plan 平面图

Rural Leisure 田园休闲

天鹅堡花园坐落在华侨城,以依山傍水、亲近自然为设计原则。一梯四户使得每个户型都南北通透,均能俯瞰庭院中心美丽而宁静的天鹅湖。 结合小区的外景设计,力求表现舒畅、自然的田园生活情趣,营造一个幽静休闲、轻松舒适的家,即使足不出户也能时刻感受大自然的清新美好。

ZHONGNAN CENTURY CITY BUILDING 5
中南世纪城 5 幢

Design company: Zhuang Zhuang Design Company **Location**: Changshu, Jiangsu **Area**: 160m²
设计公司：壮壮设计 项目地点：江苏，常熟 项目面积：160 平方米

The hostess is good at fitness and Yoga.Because of her hobby on sports,this project mixes Thai,modern and Bohemian characters to show a exotic style. Warm color is the elementary tone in the whole house, with the light color embellished.You can express your beautiful hope for life from your heart with the Yoga music.

女主人公热爱健身、瑜伽。根据其对运动的热爱，本案运寻异国浪漫情怀，结合泰式、现代以及波西米亚的混搭手法。整体以暖色为主，以浅色点缀其中的浪漫音符。随着瑜伽音乐而变换着由内而外散发的情感，抒发着对生活的美好憧憬。

Home Space Creative Design Roundups 居家空间创意集

Rural Leisure 田园休闲

图书在版编目（CIP）数据

田园休闲／深圳市海阅通文化传播有限公司主编．
北京：中国建筑工业出版社，2013.4
（居家空间创意集）
ISBN 978-7-112-15185-1

Ⅰ.①田… Ⅱ.①深… Ⅲ.①住宅—室内装饰设计—图集 Ⅳ.
①TU241-64

中国版本图书馆CIP数据核字(2013)第038826号

责任编辑：费海玲　张幼平　王雁宾
责任校对：姜小莲　陈晶晶
装帧设计：熊黎明
采　　编：李箫悦　罗　芳

居家空间创意集
田园休闲
深圳市海阅通文化传播有限公司　主编
*
中国建筑工业出版社出版、发行（北京西郊百万庄）
各地新华书店、建筑书店经销
深圳市海阅通文化传播有限公司制版
北京方嘉彩色印刷有限责任公司印刷
*
开本：880×1230毫米　1/16　印张：$5^1/_2$　字数：180千字
2013年6月第一版　2013年6月第一次印刷
定价：29.00元
ISBN 978-7-112-15185-1
　　　（23279）
版权所有　翻印必究
如有印装质量问题，可寄本社退换
（邮政编码 100037）